Abdelhafid Mimouni

Bioinorganics deciphers the vascular state

Abdelhafid Mimouni

Bioinorganics deciphers the vascular state

ScienciaScripts

Imprint

Any brand names and product names mentioned in this book are subject to trademark, brand or patent protection and are trademarks or registered trademarks of their respective holders. The use of brand names, product names, common names, trade names, product descriptions etc. even without a particular marking in this work is in no way to be construed to mean that such names may be regarded as unrestricted in respect of trademark and brand protection legislation and could thus be used by anyone.

Cover image: www.ingimage.com

This book is a translation from the original published under ISBN 978-620-6-72608-1.

Publisher:
Sciencia Scripts
is a trademark of
Dodo Books Indian Ocean Ltd. and OmniScriptum S.R.L publishing group

120 High Road, East Finchley, London, N2 9ED, United Kingdom
Str. Armeneasca 28/1, office 1, Chisinau MD-2012, Republic of Moldova, Europe
Printed at: see last page
ISBN: 978-620-8-23042-5

Bioinorganics deciphers the vascular state

Author : Dr. Abdelhafid Mimouni

An independent researcher in bioinorganic chemistry, Dr Mimouni is an expert in macromolecular synthesis and characterisation. He obtained his PhD in Chemistry from the University of Paris XII in 1997, after obtaining a Diplôme d'Études Approfondies en Systèmes Bioinorganiques from the University of Paris XI in 1993. He also obtained his Bachelor's and Master's degrees in Chemistry from the same university.

Summary: This book examines the fundamental role of bioinorganics in living systems, focusing on phenomena such as stroke and the breakdown of blood clots. It highlights the importance of metals such as zinc, iron and copper, and their interaction with various biomolecules, influencing essential processes. By exploring how these elements impact therapeutic agents, the book highlights the consequences of a metal imbalance on health. It also emphasises the need for an interdisciplinary approach between bioinorganics, chemistry and biology to meet the challenges of vascular disease. Future research, using advanced techniques such as spectroscopy, promises to shed further light on these complex interactions and pave the way for new treatment strategies. All in all, this book offers an enriching perspective on the importance of inorganic elements in maintaining health.

Book outline :

Introduction

Bioinorganics is an essential discipline that examines the role of inorganic elements, particularly metals, in biological systems. These elements play a crucial role in the regulation of physiological processes, often acting as cofactors in enzymatic reactions and thus influencing a multitude of cellular functions. Among the most notable metals are zinc, copper and iron, which are involved in mechanisms as varied as cell respiration, intracellular signalling and the immune response.

As part of haematological processes, coagulation and fibrinolysis stand out as crucial biological mechanisms. Coagulation is the process by which blood forms clots to prevent haemorrhage, while fibrinolysis is responsible for dissolving these clots once healing has begun. These two processes must be finely regulated to maintain the balance between haemostasis and tissue repair. Metals and metalloenzymes play a decisive role here. For example, calcium is essential for the activation of several coagulation factors, while zinc and copper are involved in the activity of various enzymes required for clot degradation.

The aim of this book is to explore in depth the bioinorganic mechanisms associated with coagulation and fibrinolysis, with particular emphasis on the role of metals and metalloenzymes in these processes. We will also look at the use of advanced techniques, such as XAS (X-ray absorption spectroscopy), to identify and characterise these interactions within biological systems. In doing so, we hope not only to enrich our understanding of the complex dynamics involved, but also to pave the way for new therapeutic approaches for the treatment of vascular

diseases, including stroke.

Chapter 1: Fundamental concepts of bioinorganics

1.1. Definition of bioinorganics

Bioinorganics is a scientific discipline that explores the roles of inorganic elements in biological systems. In particular, it examines how metals, as essential co-factors, participate in various biological processes. This discipline integrates inorganic chemistry concepts into biological contexts, studying the interactions between metals and biomolecules and their impact on the structure and function of biological systems (Bertini et al., 2001). It also enables us to understand how these inorganic elements influence human health and pathologies.

1.2 Role of metals and ions in biological processes

Metals such as zinc (Zn), iron (Fe), copper (Cu), manganese (Mn), vanadium (V) and cobalt (Co) play crucial roles in many biological processes.

Zinc (Zn): A Key Element for Health

Zinc is an essential trace element that plays a crucial role in many biological processes. It is involved in the structure and function of more than 300 enzymes, making it a central player in various biological mechanisms.

1. Role in Enzymatic Catalysis

Zinc is a cofactor for many enzymes, particularly those involved in :

- **Enzymatic reactions**: Zinc helps stabilise enzyme structures, facilitating essential biochemical reactions. For example, it is present in enzymes such as carboxypeptidase, which is involved in protein digestion.
- **DNA and RNA synthesis**: Zinc is crucial for enzymes such as DNA polymerase, which plays a fundamental role in DNA replication and repair.

2. Regulation of gene expression

Zinc is involved in the regulation of gene expression through :

- **Interactions with transcription factors**: Zinc contributes to the structure of zinc fingers, DNA binding motifs found in many transcription factors. These proteins regulate gene expression by binding to specific DNA sequences.
- **Response to cellular stress**: In the event of oxidative stress, zinc plays a protective role by stabilising cell membranes and helping to regulate the expression of genes involved in defence responses.

3. Impact on the immune system

Zinc is fundamental to the proper functioning of the immune system:

- **Immune cell function**: Zinc is essential for the maturation and function of T lymphocytes and macrophages, two key cell types in the immune response. Zinc deficiency can lead to a reduced immune response, increasing susceptibility to infection.

- **Cytokine production**: Zinc influences the production of cytokines, which are proteins essential for communication between immune system cells.

4. Wound healing

Zinc also plays a crucial role in the healing process:

- **Collagen synthesis**: Zinc is involved in the synthesis of collagen, a protein fundamental to tissue structure. This makes it a key element in the repair of damaged tissue.
- **Regulation of inflammation**: Zinc helps modulate the inflammatory response, promoting effective healing and reducing the risk of wound infections.

5. Dietary sources of Zinc

Zinc is present in a variety of foods, including :

- **Animal sources**: Red meat, seafood (particularly oysters) and dairy products are rich sources of zinc, with high bioavailability.
- **Plant sources**: Legumes, nuts and whole grains also contain zinc, but their bioavailability may be reduced due to the presence of phytates, which inhibit zinc absorption.

6. Deficiencies and Consequences

Zinc deficiency can have serious effects on health:

- **Immune anomalies**: A deficiency can lead to immunodeficiency, making the individual more vulnerable to infections.

- **Growth retardation**: In children, a lack of zinc can inhibit growth and development.

- **Wound healing problems**: Zinc deficiency can delay wound healing, increasing the risk of infection.

7. Overdose and toxicity

Although zinc is essential, too much can be toxic:

- **Toxicity**: Excessive zinc intake can cause nausea, vomiting and gastrointestinal problems. In the long term, over-consumption can lead to copper deficiency, as zinc interferes with its absorption.

Conclusion

Zinc is an essential trace element for enzyme catalysis, regulation of gene expression, proper functioning of the immune system and wound healing. Its balance in the diet is essential for maintaining health and preventing various diseases. Understanding the role of zinc in biological processes underlines its vital importance in nutrition and medicine.

Iron (Fe) : An Essential Element for Life

Iron is a fundamental trace element that plays crucial roles in many biological processes. Its main function is linked to oxygen transport, but it is also involved in other vital functions.

1. Oxygen transport

Iron is a key component of haemoglobin and myoglobin, two essential proteins:

- **Haemoglobin**: Found in red blood cells, haemoglobin binds to oxygen in the lungs and transports it to the body's tissues. Each haemoglobin molecule can carry up to four oxygen molecules, making iron essential for maintaining adequate oxygenation of the cells.
- **Myoglobin**: This protein, found in muscles, stores oxygen and releases it when the muscles need it, particularly during physical effort. Myoglobin is particularly important for skeletal and cardiac muscles.

2. Role in Energy Metabolism

Iron is also crucial for energy metabolism. It is an essential component of cytochromes, which are enzymes involved in the electron transport chain:

- **Cytochromes**: These enzymes play a vital role in cellular respiration by transferring electrons within the mitochondria, leading to the production of ATP, the main source of energy for cells. Iron enables the cytochromes to catalyse the reactions needed to convert chemical energy into usable energy.

3. Co-factor in Other Enzymes

In addition to its role in haemoglobin and cytochromes, iron is also a cofactor for many other enzymes:

- **Enzymes involved in DNA synthesis**: Iron is necessary for enzymes such as ribonucleotide reductases, which are essential for DNA synthesis.
- **Antioxidants**: Certain enzymes, such as catalase, which breaks down hydrogen peroxide, contain iron. This helps protect cells against oxidative stress.

4. Dietary sources of iron

Iron is available in many foods, but it comes in two main forms:

- **Haem iron**: Found in animal sources (such as red meat, fish and poultry), it is better absorbed by the body.
- **Non-heme iron**: Found in plant-based foods (such as green leafy vegetables, legumes and cereals), it is less well absorbed, but its bioavailability can be increased by vitamin C consumption.

5. Deficiencies and Consequences

An iron deficiency can have serious consequences for health:

- **Iron deficiency anaemia**: This is one of the most common forms of anaemia, characterised by a reduction in haemoglobin in the blood, leading to fatigue, weakness and reduced concentration.
- **Developmental problems**: In children, iron deficiency can affect cognitive and physical development.

6. Overdose and toxicity

Although iron is essential, too much can be toxic:

- **Iron accumulator**: Individuals with diseases such as haemochromatosis have excessive iron absorption, which can damage organs such as the liver, heart and pancreas.
- **Necessary balance**: Managing iron intake is therefore crucial, as too little or too much can have harmful effects.

Conclusion

Iron is essential for oxygen transport, energy metabolism and various enzyme functions. Its presence in the body is essential for maintaining health, and a balanced dietary intake is necessary to prevent both deficiencies and excesses. Understanding the role of iron in biological systems is essential for tackling health problems linked to its metabolism.

Copper (Cu): An Essential Metal for Enzymatic Catalysis and Health

Copper is a vital trace element that plays a multifunctional role in various biological processes. As a cofactor for several enzymes, it is essential for enzyme catalysis and the regulation of numerous physiological mechanisms.

1. Enzyme cofactor

Copper is an essential cofactor for several enzymes crucial to cell metabolism. These include :

- **Cytochrome c oxidase**: This enzyme, which plays a key role in the electron transport chain, is essential for the production of ATP in the mitochondria. Copper is involved in the transfer of electrons within this enzyme, facilitating the cellular respiratory process.
- **Other enzymes**: Copper is also necessary for the activity of other enzymes, such as lysyl oxidase, which is involved in the formation of collagen and elastin, thus contributing to the health of connective tissue.

2. Role in the Electron Transport Chain

Copper is a key component in the electron transport chain, a process fundamental to cellular respiration. In this role, it helps to :

- **Transferring electrons** : Copper facilitates the transfer of electrons between different enzyme complexes, which is essential for energy production.
- **Regulating ATP production**: Through its role in electron transfer, copper contributes to the generation of ATP, the main source of energy for cells.

3. Detoxification of Free Radicals

Copper also plays a crucial role in detoxifying free radicals, unstable molecules that can damage cells. It is involved in :

- **Anti-oxidation**: As a cofactor for enzymes such as superoxide dismutase (SOD), copper helps neutralise free radicals, protecting cells from oxidative damage.
- **Cellular protection**: By regulating levels of free radicals, copper helps to preserve cellular integrity and prevent various diseases linked to oxidative stress.

4. Role in iron metabolism

Copper is also important for iron metabolism. It helps to :

- **Facilitate iron absorption**: Copper is needed to convert ferrous iron (Fe^{2+}) into ferric iron (Fe^{3+}), which is essential for its transport in the blood by transferrin.

- **Prevent anaemia**: A copper deficiency can interfere with the absorption of iron, leading to problems such as anaemia.

5. Health implications

Adequate levels of copper are essential for maintaining good health. Copper deficiency can lead to deleterious effects, such as :

- **Neurological abnormalities**: Copper deficiencies may be associated with neurological disorders and developmental problems.
- **Reduced immunity**: Copper is involved in the proper functioning of the immune system, and a deficiency can weaken the body's defences.

6. Food sources

Copper is found in a variety of foods, including :

- **Seafood**: Oysters and other seafood are particularly rich in copper.
- **Nuts and seeds**: Nuts, especially cashew nuts and sunflower seeds, are good sources of this metal.
- **Whole grains**: Whole grains also contain significant levels of copper.

7. Beware of Excess

Although copper is essential, too much can be toxic, causing gastrointestinal disorders and liver damage. This underlines the importance of an appropriate balance in dietary intake.

Conclusion

Copper is an essential trace element that plays a variety of roles in enzyme catalysis, the electron transport chain and the detoxification of free radicals. Its presence is crucial for maintaining overall health, and particular attention must be paid to its intake to prevent deficiencies and excesses, thus ensuring that biological processes function optimally.

Manganese (Mn): An Essential Metal for Many Biological Processes

Manganese is an essential trace element that plays a crucial role in various biological processes, in particular as a cofactor for several enzymes. Its varied functions make it essential for the body to function properly.

1. Enzyme cofactor

Manganese is a key cofactor for several enzymes essential to metabolism. It participates in biochemical reactions involved in :

- **Amino acid metabolism**: Manganese is involved in transamination and deamination reactions, which are crucial for the synthesis and degradation of amino acids.
- **Carbohydrate synthesis**: It plays a role in gluconeogenesis, the process by which glucose is synthesised from non-carbohydrate precursors, essential for maintaining adequate blood glucose levels.

- **Detoxification**: Manganese is involved in the metabolism of toxins and medicines, helping to purify the body.

2. Protection against Oxidative Stress

One of the most important functions of manganese is its ability to protect cells against oxidative stress. It is a key component of superoxide dismutase (SOD), an antioxidant enzyme that catalyses the conversion of superoxide to hydrogen peroxide, thereby reducing oxidative damage in cells. This function is particularly important in tissues sensitive to damage, such as those of the brain and liver.

3. Role in mineral metabolism

Manganese is also involved in the metabolism of other essential minerals, notably calcium, phosphorus and magnesium. It plays a role in bone mineralisation, contributing to bone and joint health. Manganese deficiency can affect bone density and increase the risk of fractures.

4. Health implications

Studies have shown that adequate levels of manganese are associated with better metabolic health. Manganese deficiencies can lead to disorders such as :

- **Growth difficulties**: Particularly in children, a deficiency can affect normal development.
- **Neurological problems**: Low levels of manganese have been linked to neurological disorders, including cognitive impairment.

- **Immune dysfunction**: Given its role in metabolism and cellular protection, a deficiency can weaken the immune system.

5. Food sources

Manganese is present in a variety of foods, including :

- **Wholegrain cereals** : Rich in manganese, they are an excellent source for maintaining adequate levels.
- **Nuts and seeds**: These foods also provide a significant amount of manganese.
- **Green leafy vegetables**: Vegetables such as spinach and kale are good sources of this metal.

6. Beware of Excess

Although manganese is essential, too much can be toxic, causing neurological effects such as motor disorders similar to those seen in Parkinson's disease. This underlines the importance of an appropriate balance in dietary intake.

Conclusion

Manganese is a vital trace element involved in many biological processes, notably as an enzyme cofactor, in protection against oxidative stress and in mineral metabolism. Its presence is crucial to the maintenance of overall health, and particular attention must be paid to both its sufficient intake and balance to prevent possible adverse effects.

- **Vanadium (V): Biological Role and Potential Applications**
- Although less well known than other essential metals, vanadium plays an increasingly recognised role in a number of biological processes, including the regulation of blood sugar levels and the influence on lipid metabolism. Recent research has highlighted its potentially beneficial properties, particularly in the management of diabetes.

1. Regulation of blood sugar levels

Vanadium has been studied for its hypoglycaemic effect, i.e. its ability to lower blood glucose levels. In vitro and in vivo studies have shown that vanadium can mimic insulin, facilitating the entry of glucose into cells, which is crucial for people with type 2 diabetes. Indeed, some research has observed that vanadium can activate intracellular signalling pathways similar to those of insulin, stimulating glucose uptake by muscle and adipose tissue (Soni et al., 2012).

2. Influence on lipid metabolism

In addition to its role in regulating blood sugar levels, vanadium also appears to have an impact on lipid metabolism. Studies have suggested that it may modulate blood lipid levels, helping to reduce total cholesterol and triglycerides. This action could be beneficial for people suffering from dyslipidaemia, often associated with type 2 diabetes and other metabolic disorders.

3. Mechanisms of action

The mechanisms of action of vanadium in the human body are still being explored. It is proposed that vanadium interacts with various enzymes and proteins involved in glucose and lipid metabolism. For example, it could inhibit key enzymes such as glycerol-3-phosphate dehydrogenase, thereby influencing lipid synthesis and glucose breakdown.

4. Therapeutic perspectives

Vanadium's properties make it a potential candidate for the development of anti-diabetic treatments. Although clinical studies are still needed to establish its efficacy and safety, vanadium-based compounds such as vanadyl sulphate are currently being evaluated. These studies could lead to new approaches to diabetes management, particularly for patients who do not respond well to conventional treatments.

5. Precautions and considerations

Despite its promising effects, the use of vanadium as a supplement should be approached with caution. High doses can be toxic and lead to undesirable effects, such as gastrointestinal disorders or negative impacts on renal function. It is therefore essential to conduct in-depth research to better understand the therapeutic window of vanadium and to establish clear guidelines for its use.

- **Conclusion**
- Vanadium, although less frequently mentioned than other trace elements, deserves increased attention because of its potential roles in the regulation of blood glucose and lipid metabolism. With continued research, it could prove to be a key element in the development of new therapeutic strategies for the treatment of diabetes and other metabolic disorders.

Cobalt (Co): Essential role in metabolic health

Cobalt, although often less prominent than other trace elements, is an essential metal that plays a crucial role in a number of biological processes. Its most notable contribution is its function as a key constituent of vitamin B12, also known as cobalamin, which is essential for human health.

1. Synthesis of Vitamin B12

Cobalt is a central element in the molecular structure of vitamin B12. This vitamin is essential for the formation of red blood cells and the proper functioning of the nervous system. Without sufficient cobalt, B12 synthesis is compromised, which can lead to megaloblastic anaemia, characterised by the production of abnormally large and poorly functioning red blood cells.

2. Red blood cell formation

Vitamin B12 is required for DNA synthesis in haematopoietic stem cells, which is crucial for the production of red blood cells. A vitamin B12 deficiency, often linked to a cobalt deficiency, can lead to a reduction in red blood cell production, causing fatigue, weakness and other symptoms associated with anaemia. Cobalt is therefore essential not only for the production of vitamin B12, but also for the overall health of the circulatory system.

3. Cellular metabolism

Cobalt also plays a role in various metabolic processes within cells. It is involved in the synthesis of certain enzymes and coenzymes, influencing key metabolic pathways, particularly those linked to energy production. In addition, cobalt helps regulate homocysteine, an amino acid whose high levels are associated with an increased risk of cardiovascular disease.

4. Neurological health

Cobalt, through vitamin B12, is crucial for maintaining neurological health. It is involved in the formation of myelin, the substance that covers and protects the nerves. A vitamin B12 deficiency can lead to neuropathy, memory problems and other neurological dysfunctions. An adequate intake of cobalt is therefore essential to prevent cognitive and neurological problems.

5. Immune function

Cobalt and vitamin B12 also play a role in the proper functioning of the immune system. A deficiency in B12 can alter the immune response, making the body more vulnerable to infection. Cobalt therefore contributes indirectly to the body's defence against various diseases.

6. Perspectives and Considerations

Although cobalt is essential, it is important to maintain an appropriate balance, as excessive levels can be toxic. Occupational or environmental exposures can lead to adverse effects, including respiratory disorders and skin problems. It is therefore crucial to monitor cobalt intake, particularly for people on diets low in vitamin B12 or suffering from malabsorption.

Conclusion

Cobalt is a vital trace element that plays a fundamental role in vitamin B12 synthesis, red blood cell formation, cell metabolism, neurological health and immune function. Its presence in the body is essential to ensure healthy and efficient biological processes, underlining the importance of adequate intake through diet or supplements when necessary.

These metals act as cofactors, promoting biochemical reactions essential to life and helping to regulate numerous physiological processes.

1.3 Interaction of metal complexes with biomolecules

Metal complexes interact dynamically with various biomolecules, influencing their structures and functions. These interactions can modify the physicochemical properties of biomolecules, affecting processes such as enzyme catalysis, cell signalling and the stability of protein structures (López et al., 2013).

For example, copper complexes can induce conformational changes in proteins, altering their enzymatic activity. Similarly, the interactions of metal ions with DNA can influence genetic regulation and DNA repair (Bennett et al., 2009).

Manganese complexes, for example, can act as cofactors for antioxidant enzymes, protecting cells from oxidative stress. In addition, vanadium complexes can interact with insulin signalling pathways, highlighting their potential in the treatment of metabolic disorders.

These complex interactions are at the heart of our understanding of bioinorganic mechanisms and their relevance to cellular biology. In-depth study of these relationships between metals and biomolecules offers prospects for the development of innovative therapies, as well as a better understanding of diseases linked to metal imbalances.

Chapter 2: Blood coagulation

2.1. Mechanisms of coagulation

Blood coagulation is a fundamental and complex process that stops bleeding and restores the integrity of blood vessels. This mechanism is based on a series of enzymatic cascade reactions, often classified into three main pathways: the extrinsic pathway, the intrinsic pathway and the common pathway.

1. Extrinsic pathway

The extrinsic pathway is the first response to tissue damage. It is activated by the release of tissue factor (TF), a protein found outside blood vessels. When a vessel is damaged, TF comes into contact with the blood and, in the presence of calcium, activates factor VII. This activated factor VII (VIIa) then catalyses the conversion of factor X into factor Xa. This step is crucial, as factor Xa constitutes a convergence point for the subsequent coagulation pathways.

2. Intrinsic pathway

The intrinsic pathway is triggered by blood coming into contact with foreign surfaces, such as damaged tissue or medical devices. It involves several coagulation factors, notably factor XII, which activates factor XI, followed by activation of factor IX by factor XI. This activation cascade also leads to the generation of factor Xa, reinforcing clot formation.

3. Voie Commune

The two coagulation pathways come together in the common pathway. Here, factor Xa, in collaboration with factor Va, catalyses the conversion of prothrombin (factor II) into thrombin (factor IIa). Thrombin plays a central role in coagulation, as it converts fibrinogen, a soluble protein, into fibrin, which forms an insoluble network. This fibrin network stabilises the blood clot, preventing blood loss and promoting healing of damaged tissue.

4. Regulation of coagulation

The regulation of coagulation is crucial to avoid complications such as thrombosis, which can lead to serious vascular accidents. Natural inhibitors, such as antithrombin, protein C and protein S, modulate the activity of coagulation factors. For example, antithrombin inhibits activated coagulation factors, while protein C, activated by thrombin, degrades factors Va and VIIIa. This regulation ensures a delicate balance between coagulation and fibrinolysis, guaranteeing that the coagulation process occurs in a controlled and efficient manner (Hoffman et al., 2009).

Conclusion

Understanding coagulation mechanisms, including extrinsic, intrinsic and common pathways, is essential for the diagnosis and treatment of coagulation disorders. In-depth knowledge of these processes can also lead to the development of innovative treatments aimed at preventing thromboembolic complications and improving the management of patients with bleeding disorders.

2.2. Role of calcium ions and other metals

Calcium (Ca^{2+}) and other metals play crucial roles in the blood coagulation process. Their presence is essential to ensure the proper functioning of the enzymatic cascades that help prevent haemorrhage and restore vascular integrity after injury.

1. Role of Calcium

Activation of coagulation factors

Calcium is essential for the activation of many coagulation factors. It acts as a cofactor at several stages in the enzymatic cascades that regulate coagulation. For example, it facilitates the interaction between coagulation factors and cell membranes, which is crucial for their activation. Without calcium, coagulation factors, such as factors II, VII, IX and X, cannot effectively bind to membranes, preventing the activation process and potentially compromising coagulation (Schwartz et al., 2006).

Platelet aggregation

Calcium also plays a key role in platelet aggregation. When a vascular lesion occurs, blood platelets become activated and begin to stick together to form a clot. Calcium is required for the release of platelet granules containing pro-coagulant substances such as ADP and thromboxane A2. These substances encourage platelet aggregation, thereby amplifying clot formation and stabilising the area of injury.

2. Role of other Metals

Zinc

Zinc is another essential metal that influences coagulation. It participates in the activation of certain enzymes involved in the coagulation process, such as metalloproteinases. In addition, zinc plays a role in platelet function and can modulate the inflammatory response, which is important in the context of vascular lesions.

Magnesium

Magnesium also has significant effects on coagulation. It acts as a stabiliser of protein complexes and is involved in the activation of various coagulation factors. In particular, it is required for the conversion of prothrombin to thrombin, a key step in the coagulation cascade. A good magnesium balance is crucial for maintaining effective coagulation and avoiding complications such as thrombosis.

3. Interactions and Balance

The interactions between calcium, zinc, magnesium and other metals are essential for the proper functioning of the coagulation pathways. An imbalance in the levels of these metals can lead to coagulation disorders. For example, a calcium deficiency can reduce the activation of coagulation factors, while an excess of calcium can favour thrombotic events.

Conclusion

Calcium and other metals such as zinc and magnesium are essential to the coagulation process. Their role in the activation of coagulation factors and platelet aggregation highlights the importance of an adequate mineral balance in maintaining blood homeostasis. An in-depth understanding of these mechanisms may contribute to the development of new therapeutic approaches for treating coagulation disorders and preventing vascular disease.

2.3 Understanding Stroke and Fibrinolysis

Stroke is a major medical emergency, characterised by a sudden interruption of the blood supply to the brain. This interruption can lead to irreversible brain damage, permanent neurological deficits and even death. Strokes fall into two main categories, each with distinct mechanisms and treatments.

1. Types of stroke

Ischaemic stroke

Ischaemic stroke accounts for around 85% of all strokes. It is caused by a blood clot that blocks a cerebral vessel, reducing the blood supply to a specific area of the brain. This blockage can result from a number of factors, such as atherosclerosis, where fatty deposits accumulate in the arteries, or the formation of a thrombus following atrial fibrillation. Brain cells deprived of oxygen and nutrients begin to die within minutes of the blockage, causing permanent damage.

Haemorrhagic stroke

Haemorrhagic stroke, which accounts for around 15% of strokes, occurs when a ruptured blood vessel causes bleeding in the brain. This can be caused by uncontrolled high blood pressure, vascular malformations or trauma. Bleeding causes increased pressure on brain tissue and can also lead to secondary ischaemia.

2. Role of Fibrinolysis

Fibrinolysis is the process by which the body breaks down fibrin, a protein essential for the formation of blood clots. This mechanism is crucial for re-establishing blood circulation after a clot has formed, thereby limiting brain damage.

Mechanism of Fibrinolysis

Activation

- **Plasminogen conversion**: Fibrinolysis begins with the conversion of plasminogen, a precursor protein, into plasmin, the active enzyme that breaks down fibrin. This conversion is a key stage in the dissolution of clots.

- **Role of Tissue Plasminogen Activator (tPA)**: tPA catalyses the conversion of plasminogen to plasmin. This enzyme is released by endothelial cells in response to vascular injury and plays a central role in regulating fibrinolysis.

Control

- **Fibrinolysis inhibitors**: To avoid complications such as haemorrhage, fibrinolysis is strictly regulated. The plasmin inhibitor (PAI-1) is a key example, as it prevents tPA from converting plasminogen into plasmin.

- **Balance between pro-fibrinolytics and inhibitors**: A balance between pro-fibrinolytic agents and their inhibitors is essential. Excessive inhibition can lead to thrombosis, while overactive fibrinolysis can cause bleeding.

3. Stimulation of fibrinolysis

Medicines

Thrombolytic agents such as rtPA (recombinant tissue plasminogen activator) are used to stimulate fibrinolysis in ischaemic stroke. Rapid administration of rtPA can restore blood flow by rapidly converting plasminogen to plasmin, but to be effective, this treatment must be administered within 3 to 4.5 hours of the onset of symptoms.

Physical activity

Regular exercise improves blood circulation and reinforces the natural mechanisms of fibrinolysis. Moderate physical activity promotes cardiovascular health by increasing blood circulation and improving the activation of fibrinolysis.

Power supply

A diet rich in omega-3 fatty acids and antioxidants also contributes to vascular health. Omega-3 fatty acids, found in foods such as oily fish and nuts, can reduce inflammation and promote optimal blood circulation, while antioxidants help protect endothelial cells from damage.

Conclusion

Understanding the mechanisms of fibrinolysis and its role in the treatment of stroke is essential for the development of effective therapeutic strategies. Stroke management requires an integrated approach, including medical interventions, lifestyle modifications and nutritional recommendations to optimise vascular health and reduce the risk of stroke.

2.4. Balance between coagulation and fibrinolysis

The balance between coagulation and fibrinolysis is essential for maintaining homeostasis in the circulatory system. This dynamic process ensures that the body can quickly form a clot to stop a haemorrhage, while also having the ability to break down the clot once the blood vessel has been repaired, preventing excessive blockage.

The importance of balance

Excessive coagulation can lead to thrombosis, i.e. the formation of clots in blood vessels that can obstruct blood flow, causing complications such as ischaemic stroke, myocardial infarction or pulmonary embolism. On the other hand, inadequate fibrinolysis can lead to haemorrhage, where the body is unable to maintain a clot long enough to stop bleeding.

Consequences of an imbalance

Various factors can upset this delicate balance:

- **Age**: With age, coagulation mechanisms can become less effective, leading to a predisposition to haemorrhage. At the same time, changes in blood composition and platelet function can increase the risk of thrombosis.

- **Diet**: An unbalanced diet can influence coagulation and fibrinolysis. For example, a zinc deficiency can alter platelet function, hampering coagulation, while an over-consumption of calcium can encourage thromboembolic events. Similarly, a diet rich in omega-3 fatty acids can promote healthy fibrinolysis by reducing inflammation and thinning the blood.

- **Pathologies**: Certain medical conditions can affect the balance between coagulation and fibrinolysis. Diseases such as hyperlipidaemia, diabetes and hereditary coagulation disorders can predispose to either thrombosis or haemorrhage.

Regulatory mechanisms

Several mechanisms regulate the balance between coagulation and fibrinolysis:

- **Coagulation factors and inhibitors**: Coagulation factors and their natural inhibitors, such as antithrombin and protein C, work in tandem to maintain this balance. Antithrombin, for example, inhibits activated coagulation factors, while protein C inactivates factors Va and VIIIa, thereby reducing thrombin formation.

- **Balance between pro-fibrinolytics and inhibitors**: Fibrinolytic agents, such as tPA, are counterbalanced by inhibitors such as PAI-1. An imbalance in this dynamic can favour clot formation or lead to excessive degradation of existing clots.

Clinical implications

Understanding the mechanisms underlying this balance is crucial to the development of innovative treatments to prevent and treat vascular diseases, particularly stroke. For example:

- **Targeted therapies**: Drugs can be developed to specifically modulate coagulation or fibrinolysis, allowing treatments to be tailored to the patient's risk profile.
- **Nutritional interventions**: Dietary recommendations can be established to promote a healthy balance between coagulation and fibrinolysis, incorporating key nutrients such as zinc and omega-3 fatty acids.
- **Monitoring risk factors**: Rigorous monitoring of the risk factors associated with imbalances can help identify high-risk patients and intervene preventively.

In short, maintaining a balance between coagulation and fibrinolysis is fundamental to vascular health. A thorough understanding of this regulatory system is essential for the development of effective strategies to prevent and treat vascular disease, thereby improving clinical outcomes for patients.

Chapter 3: Fibrinolysis and its mechanisms

3.1. Fibrinolysis process

Fibrinolysis is the biological process by which blood clots are broken down, allowing normal blood flow to be restored. This mechanism is essential after coagulation to avoid prolonged obstruction of blood vessels. The fibrinolysis process takes place in several key stages:

1. **Formation of fibrin**: After coagulation, fibrinogen, a soluble protein in plasma, is converted into fibrin by the action of thrombin. Fibrin forms an insoluble network that stabilises the clot.

2. **Plasminogen activation**: Plasminogen, an inactive enzyme, is incorporated into the clot. It is converted to plasmin, the key enzyme in fibrinolysis, mainly by the action of tissue plasminogen activator (tPA) and urokinase (uPA) (Baker et al., 2012).

3. **Fibrin degradation**: Plasmin degrades fibrin into fibrin degradation products, such as D-dimers, which are important markers of fibrinolytic activity (Mackman et al., 2009).

4. **Clot resorption**: Once the fibrin has been broken down, the clot is resorbed, allowing the blood vessel to return to its normal state.

3.2 Key enzymes: Plasminogen and Plasmin

Plasminogen and plasmin play central roles in the fibrinolysis process:

- **Plasminogen**: Synthesised mainly by the liver, plasminogen is an inactive precursor which must be activated to take part in fibrinolysis. It binds to fibrin, which facilitates its activation (Björk et al., 2008).

- **Plasmin**: This is the active enzyme that breaks down fibrin into soluble fragments. Its formation is regulated by inhibitors, such as plasmin inhibitor (PAI-1), which prevent excessive clot degradation (Nielsen et al., 2010). Regulation of plasmin activity is crucial to prevent excessive bleeding.

3.3 Influence of metals on enzyme activity

Metals, as essential cofactors, influence the activity of enzymes involved in fibrinolysis:

- **Calcium (Ca^{2+})**: Calcium is crucial for plasminogen activation. Its presence promotes the binding of plasminogen to cell surfaces and fibrin, facilitating its conversion to plasmin (Hultin et al., 2004).

- **Zinc (Zn^{2+})**: Zinc plays a stabilising role for many enzymes, including some involved in fibrinolysis. Studies show that adequate concentrations of zinc can enhance plasmin activity, while excess or deficiency can inhibit its function (Jiang et al., 2011).

- **Copper (Cu^{2+})**: Although less directly involved, copper also influences the structure and function of fibrinolytic proteins through its effects on the oxidation and post-translational modification of enzymes (Mann et al., 2006).

Understanding the effects of metals on fibrinolysis opens up prospects for the development of new therapies and treatments targeting these mechanisms.

Chapter 4: Thrombolytic agents and bioinorganics

4.1. Mechanisms of Action of Thrombolytic Agents

Thrombolytic agents play an essential role in the management of medical emergencies, particularly in the treatment of ischaemic stroke and myocardial infarction. Their effectiveness is based on their ability to dissolve blood clots by initiating a complex process to break down fibrin, a key clot-forming protein. Thrombolytic agents act primarily by converting plasminogen, an inactivated protein, into plasmin, the enzyme responsible for breaking down fibrin. Here is an overview of the main classes of thrombolytic agents and their mechanisms of action:

1. Tissue Plasminogen Activators (tPA)

Tissue plasminogen activators, such as alteplase, are among the most widely used thrombolytic agents in clinical practice. Their mechanism of action is as follows:

- **Targeting Fibrin-Bound Plasminogen**: tPA has a specific affinity for plasminogen when bound to fibrin, the main clot-forming protein. By binding to this complex, tPA activates plasminogen to plasmin in a localised manner, maximising clot degradation while minimising systemic activation of fibrinolysis (Peters et al., 2014).
- **Clot dissolution**: Once converted to plasmin, the enzyme breaks down fibrin into soluble degradation products, leading to clot dissolution. This process is particularly crucial in ischaemic stroke, where time is of the essence in reducing brain damage.

2. Urokinase

Urokinase is another thrombolytic enzyme, but it has distinct characteristics:

- **Plasminogen activation**: Unlike tPA, urokinase activates plasminogen less selectively. It works by binding to circulating plasminogen and converting it to plasmin, but its ability to target fibrin is less marked, which can lead to side effects associated with systemic activation (Greish et al., 2009).
- **Clinical use**: Although less commonly used than tPA, urokinase can be beneficial in specific contexts, such as in the treatment of pulmonary embolism or certain cases of ischaemia.

3. Streptokinase

Streptokinase is another class of thrombolytic agents:

- **Plasminogen Activation Induction**: It binds to plasminogen, forming a complex that induces the conversion of plasminogen to plasmin. This process is less specific than tissue activators, which can lead to more diffuse and less controlled fibrinolysis (Kumar et al., 2015).
- **Effectiveness in various contexts**: Although less targeted, streptokinase is effective in certain clinical situations, notably myocardial infarction, where its ability to rapidly dissolve clots can save lives.

4. Challenges and Clinical Considerations

The use of thrombolytic agents is crucial, but comes with challenges:

- **Time of administration**: The efficacy of thrombolytic agents is highly dependent on the time of administration. Studies show that administration within three to four hours of symptom onset can significantly improve clinical outcomes. Delay may reduce their efficacy and increase the risk of complications (Sullivan et al., 2017).
- **Specificity and Side Effects**: The specificity of thrombolytic agents for plasminogen is a key factor influencing their safety. Non-targeted activation can lead to haemorrhage, underlining the importance of appropriate selection of the type of agent to be used according to the patient's clinical condition.

5. Future prospects

Research into thrombolytic agents continues to evolve, with efforts to :

- **Improving selectivity**: The design of new agents that could offer more targeted plasminogen activation and minimise adverse effects.
- **Therapeutic combinations**: Exploring the use of thrombolytic agents in combination with antiplatelet agents or other treatments to increase efficacy while reducing the risk of haemorrhage.

Conclusion

Thrombolytic agents are essential tools in the management of cardiovascular emergencies, and their mechanism of action relies on the precise conversion of plasminogen to plasmin. By better understanding the differences between the various classes of agents, their specificity, and the associated clinical challenges, it will be possible to optimise their use and improve patient outcomes. Research continues to play a crucial role in developing new strategies to maximise the efficacy of thrombolytic treatments.

4.2 Thrombolysis and Metal Complexes

Thrombolysis, the process of dissolving blood clots, is a crucial therapeutic intervention in the treatment of ischaemic stroke and myocardial infarction. Metal complexes, often considered as simple cofactors, play a significant role in modulating the efficacy of thrombolytic agents through several mechanisms.

Enzyme stabilisation

Metal complexes play a crucial role in the stabilisation of thrombolytic enzymes, which can significantly enhance their activity and efficacy. Here is a more in-depth development of this mechanism, focusing on specific metal ions and their therapeutic implications:

1. Zinc ions (Zn^{2+})

Zinc ions are notable for their ability to stabilise certain thrombolytic enzymes, such as plasmin and plasminogen activator tissue (tPA).

- **Stabilisation of plasmin**: Plasmin is the key enzyme involved in the breakdown of fibrin, the protein that forms blood clots. Studies have shown that the addition of Zn^{2+} ions improves the conformation and stability of plasmin, enabling it to function more effectively under physiological conditions. This stabilisation helps to prolong the enzyme's duration of action, facilitating faster and more efficient clot dissolution (Khan et al., 2016).

- **Stabilisation of tPA**: Similarly, tPA, which activates plasminogen to plasmin, also benefits from the presence of zinc. By stabilising this enzyme, zinc enables faster, more targeted activation of plasminogen, increasing the efficiency of the fibrinolysis process.

2. Implications for Treatments

Integrating zinc complexes into thrombolytic drug formulations opens up interesting prospects for optimising treatments.

- **Improved formulations**: By adding zinc complexes to thrombolytic drugs, it is possible to increase their stability and efficacy. For example, formulations containing zinc ions could prolong the activity of thrombolytic enzymes, enabling them to have a longer-lasting action in the body.

- **New developments**: This approach could also inspire the creation of new classes of drugs that exploit enzymatic stabilisation by metal complexes.

Such treatments could be particularly valuable for patients requiring rapid and effective intervention in cardiovascular emergencies.

3. Future research

Research should continue to explore the mechanisms by which metal complexes, particularly zinc, influence the activity of thrombolytic enzymes:

- **Mechanical studies**: In-depth studies into the molecular mechanisms by which metal ions stabilise these enzymes could shed light on ways of optimising treatments.
- **Clinical trials**: The use of zinc complexes in clinical trials could provide valuable data on their efficacy in the treatment of ischaemic stroke and other blood clot-related conditions.

Conclusion

The stabilisation of thrombolytic enzymes by metal complexes, particularly zinc ions, represents a promising strategy for improving the efficacy of thrombolytic treatments. By incorporating these complexes into drug formulations, it is possible to optimise clinical results and provide more effective treatments for the management of cardiovascular emergencies. Future research should focus on understanding the underlying mechanisms and on the clinical application of these innovative approaches.

Influence on Structure

Interactions between metals and thrombolytic enzymes can profoundly influence their structure and, consequently, their efficacy. This section looks at these interactions, focusing on copper complexes and their clinical implications.

1. Copper complexes (Cu^{2+})

Copper complexes are known to interact with thrombolytic enzymes, particularly plasmin, which can modify their functional activity.

- **Inhibition of plasmin activity**: A study has shown that copper ions (Cu^{2+}) can bind to the active sites of plasmin, inhibiting its ability to degrade fibrin. This interaction not only alters the conformation of the enzyme, but also reduces its efficiency in the fibrinolysis process (Cohen et al., 2013). As a result, the presence of copper can compromise clot dissolution, making thrombolytic treatments less effective.

2. Clinical consequences

The influence of metals, particularly copper complexes, on the enzymatic activity of thrombolytic agents raises major concerns for clinical practice.

- **Risks of Unintentional Inhibition**: Understanding these interactions is crucial, as unintentional inhibition of thrombolytic enzymes could have serious consequences for clinical outcomes. For example, the use of drugs containing metals such as copper could impair treatment efficacy, increasing the risk of treatment failure in cardiovascular emergencies.

- **Evaluation of drug composition**: This issue highlights the importance of careful evaluation of drug formulations administered to patients. Clinicians need to be aware of potential interactions between metals and thrombolytic enzymes, and further studies are needed to determine the synergistic or antagonistic effects of metal complexes in thrombolytic therapy.

3. Research prospects

Future research is essential to deepen our understanding of the interactions between metals and thrombolytic enzymes:

- **Mechanical and structural studies**: Mechanical studies could examine how copper complexes and other metals modify the three-dimensional structure of enzymes, thereby impacting their activity. The use of techniques such as NMR spectroscopy or X-ray crystallography could provide valuable information on these interactions.

- **Drug development**: Based on the results of these studies, it may be possible to design optimised thrombolytic drugs that minimise the inhibitory effects of undesirable metals while maximising the efficacy of thrombolytic enzymes.

Conclusion

The influence of metals, particularly copper complexes, on the structure and activity of thrombolytic enzymes has significant clinical implications. Understanding these interactions is crucial to ensuring the efficacy of thrombolytic treatments, requiring rigorous evaluation of drug composition and ongoing research to optimise therapeutic strategies.

3. Modulation of signalling pathways

Metal complexes also influence the cellular signalling pathways that regulate fibrinolysis. This is mainly through their interactions with specific receptors on platelets:

- **Interactions with platelet receptors**: Certain metals can bind to receptors on platelets, altering their activation and aggregation. For example, calcium (Ca^{2+}) and magnesium (Mg^{2+}) ions are known to play a crucial role in platelet activation. By modulating these signals, metal complexes can influence the initiation and regulation of fibrinolysis (Rosales et al., 2017).

- **Therapeutic strategies**: By manipulating these signalling pathways using metal complexes, it may be possible to develop targeted therapies that improve the thrombolytic response by promoting more rapid and effective clot dissolution.

4. Future prospects

The interaction between thrombolytic agents and metal complexes opens up new avenues for the development of innovative treatments.

- **Research into new metal complexes**: Exploring new metal complexes and their effects on thrombolytic enzymes could lead to the discovery of compounds with improved efficacy.
- **Optimising formulations**: By incorporating metal complexes into thrombolytic drug formulations, it will be possible to improve not only their stability, but also their specificity and efficacy, making treatments safer and more effective.

Conclusion

Metal complexes play an underestimated but crucial role in modulating the efficacy of thrombolytic agents. Through their ability to stabilise enzymes, influence their structure and modulate signalling pathways, these metals open the way to new therapeutic strategies. An in-depth understanding of these interactions could transform the way thrombolytic treatments are developed and administered, with significant implications for the management of thromboembolic disorders.

4.3. Development of New Agents Based on Inorganic Chemistry

Innovation in the field of thrombolytic agents is booming, and inorganic chemistry is playing a central role in the development of new therapeutic strategies. Here's a look at some promising approaches that could transform the treatment of thromboembolism.

1. Metallic nanoparticles

The use of nanoparticles, such as those based on gold or silver, represents a significant advance in the delivery of thrombolytic agents.

- **Improved Bioavailability**: Nanoparticles can encapsulate thrombolytic agents, protecting them from premature degradation in the bloodstream. This can increase their bioavailability, enabling a more prolonged and targeted action (Singh et al., 2018).
- **Reducing side effects**: By specifically targeting sites of action in the body, nanoparticles can minimise the side effects associated with systemic administration of drugs. This could lead to safer and more effective treatments.

2. Coordination complexes

Coordination complexes, particularly those based on transition metals, are another promising avenue for the development of thrombolytic agents.

- **Specific affinity for plasminogen**: These complexes can be designed to interact specifically with plasminogen, increasing their efficiency in converting plasminogen into plasmin. One study showed that coordination complexes could be optimised to activate plasminogen in a more targeted manner, thereby improving clot dissolution (Zhang et al., 2020).

- **Optimising Release**: By integrating modulatable ligands into coordination complexes, it is possible to control the release of thrombolytic agents, which could maximise their efficacy while reducing the risk of adverse effects.

3. Combination therapies

Combining thrombolytic agents with metal complexes opens the way to new therapeutic approaches.

- **Multi-faceted targeting**: By combining thrombolytic agents with metal complexes, it becomes possible to target several aspects of coagulation and fibrinolysis simultaneously. For example, an agent may act to dissolve the clot, while a metal complex could modulate the signalling pathways involved in coagulation (Li et al., 2021).
- **Synergistic approaches**: These combined therapies could offer synergistic effects, improving the overall efficacy of thrombolytic treatments. Current research is focusing on identifying the best combinations and the mechanisms underlying their efficacy.

Conclusion

The development of new thrombolytic agents based on inorganic chemistry is a promising and rapidly evolving field. The exploration of metal nanoparticles, coordination complexes and combined therapies highlights the importance of an interdisciplinary approach to improving thrombolytic treatments. By integrating the principles of inorganic chemistry into medicine, it is possible to design innovative solutions to combat thromboembolism, offering new prospects for patients suffering from cardiovascular conditions.

Chapter 5: Metals and Vascular Diseases

5.1. Role of trace elements (Cu, Zn, Fe) in vascular health

Trace elements such as copper (Cu), zinc (Zn) and iron (Fe) play essential roles in vascular health, although they are only present in very low concentrations in the body. Their importance lies in their participation in various biological functions that contribute to the integrity and function of vascular systems.

Copper (Cu)

Copper is a vital trace element that influences a number of physiological processes linked to vascular health:

- **Collagen formation**: Copper is essential for the synthesis of collagen, an essential structural protein that gives strength and elasticity to blood vessels. As a cofactor of lysyl oxidase, it facilitates the formation of cross-links in collagen and elastin molecules, crucial for maintaining vascular integrity (Kühn et al., 2017).
- **Risk of vascular fragility**: A copper deficiency can compromise the proper formation of collagen, leading to vascular fragility. This can increase the risk of blood vessel rupture, which has serious implications, particularly in the context of cardiovascular disease.
- **Natural antioxidant**: Copper also plays a role as an antioxidant by participating in the dismutation of free radicals, thus helping to protect vascular cells against oxidative stress.

Zinc (Zn)

Zinc is a multifunctional trace element whose effects on vascular health are particularly noteworthy:

- **Immune regulation**: As a regulator of the immune response, zinc helps to modulate inflammatory responses. Chronic inflammation is a major risk factor for cardiovascular disease.

- **Muscle contraction**: Zinc is involved in regulating vascular smooth muscle contraction. Adequate zinc availability is essential to maintain vascular tone and prevent conditions such as hypertension (Maret, 2018).

- **Wound healing**: This metal also plays a crucial role in wound healing, a process that depends on vascular health. Adequate levels of zinc are associated with a reduced risk of vascular complications, helping to prevent cardiovascular disease.

Iron (Fe)

Iron is an essential metal whose role in vascular health is twofold, being both vital and potentially dangerous:

- **Oxygen transport**: Iron is a key component of haemoglobin, enabling oxygen to be transported in the blood. This process is fundamental to cellular respiration and energy production, directly affecting tissue function, including that of blood vessels.

- **Oxidative stress**: However, excess iron in the body can lead to the formation of free radicals, inducing oxidative stress. This phenomenon is associated with cell damage and atherosclerosis, one of the main contributors to cardiovascular disease (Miller et al., 2015).

- **Balancing iron levels**: It is crucial to maintain a balance between iron levels, as both deficiency and overload can adversely affect vascular health. Proper iron management is therefore essential to prevent the complications associated with cardiovascular disease.

Conclusion

Trace elements such as copper, zinc and iron are fundamental to vascular health. Their role in a variety of biological processes, from the formation of solid vascular structures to the regulation of immune and oxidative responses, underlines the importance of maintaining an adequate balance of these metals in the body. A better understanding of their impact may contribute to more effective prevention and treatment strategies for vascular disease.

5.2 Impact of Metal Deficiency or Excess

Metal imbalances in the body, whether due to deficiency or excess, can have serious consequences for vascular health. These imbalances influence not only endothelial function, but also the regulation of inflammatory and oxidative processes.

Trace element deficiency

1. Zinc deficiency: Zinc deficiency is associated with increased inflammation and endothelial dysfunction. Zinc plays a crucial role in regulating the immune response and protecting against oxidative stress.

- **Inflammation and atherosclerosis**: A study by Bach et al (2020) showed that insufficient levels of zinc increased inflammatory markers in the blood, thereby promoting the development of atherosclerosis. This is due to the inability of the vascular endothelium to regenerate properly, resulting in a failure of blood vessel repair mechanisms.

2. Copper deficiency : Copper is essential for collagen formation and vascular health. Copper deficiency can lead to reduced vascular elasticity and tissue damage.

- **Impact on elasticity**: Copper is involved in the synthesis of lysyl oxidase, an enzyme essential for the formation of cross-links in collagen and elastin. A deficiency can therefore cause vascular rigidity, increasing the risk of hypertension and other cardiovascular complications.

3. Iron deficiency: Although iron is necessary for oxygen transport, a deficiency can also affect vascular health.

- **Erythrocyte dysfunction**: Iron deficiency can lead to anaemia, which reduces the oxygen-carrying capacity of the blood, affecting cardiac function and tissue perfusion.

Excess metals

1. Iron overload: An excess of iron, as in haemochromatosis, can cause damage to tissues, including blood vessels.

- **Oxidative stress and free radicals**: Excess iron promotes the formation of free radicals, leading to oxidative stress that can damage endothelial cells and cause chronic inflammation. A study by Boyer et al (2016) showed that patients suffering from haemochromatosis had an increased risk of stroke, illustrating the deleterious effects of iron overload on vascular health.

2. Effects of other metals: Other metals, such as manganese and copper, can also have negative effects when present in excess.

- **Copper toxicity**: An excess of copper is associated with neurological disorders and toxic effects on tissues, including blood vessels. This toxicity can disrupt normal vascular functions.

General imbalance

The complex interaction between these metals and their impact on vascular health highlights the importance of a precise balance.

- **Metal Interaction**: Excess iron can inhibit zinc absorption, exacerbating zinc deficiency and affecting vascular health (Zhou et al., 2019). This interrelationship highlights the need to consider the levels of several metals simultaneously, rather than focusing on a single element.

Conclusion

Metal imbalances in the body, whether deficiencies or excesses, can seriously compromise vascular health. A better understanding of these interactions is essential if we are to develop effective prevention and treatment strategies to maintain an optimal metal balance for cardiovascular health.

5.3. Case Studies: Stroke and Metal Imbalances

Metal imbalances play a crucial role in the development of stroke. Here are several case studies that illustrate how deficiencies or excesses of certain metals can contribute to an increased risk of stroke.

Zinc deficiency and stroke

A study by Ravindran et al (2020) revealed a significant correlation between zinc deficiency and an increase in cerebral ischaemic events, particularly in elderly subjects.

- **Observation of Zinc Levels**: Researchers measured zinc levels in the plasma and tissues of patients who had suffered a stroke. The results showed that zinc levels were significantly lower in these patients compared with those who had no ischaemic events.

- **Underlying mechanisms** : Zinc is involved in a number of biological processes, including protection against oxidative stress and inflammation, two key factors in the development of stroke. Zinc deficiency could compromise these defence mechanisms, making the brain more vulnerable to ischaemic damage.

Haemochromatosis and stroke

Haemochromatosis, a condition characterised by iron overload, is also associated with an increased risk of stroke. A study by Cohen et al (2019) highlighted this link.

- **Increased risks**: Patients with haemochromatosis had twice the risk of stroke as the general population. This increased risk is attributed to the pro-oxidant effects of iron, which can cause damage to endothelial cells and promote the formation of blood clots.

- **Consequences of Iron Accumulation**: Iron overload can lead to chronic inflammation and oxidative stress, worsening vascular conditions and increasing the likelihood of ischaemic events.

Copper deficiency and vascular complications

Copper levels in the body are also essential for cardiovascular health. Studies show that patients with copper deficiency often suffer from vascular complications.

- **Link to Vascular Elasticity**: Harris et al (2018) showed that insufficient levels of copper were associated with reduced elasticity of blood vessels. This can contribute to hypertension and other cardiovascular problems, increasing the risk of stroke.

- **Role of copper in vascular health**: Copper is essential for the formation of collagen and elastin, proteins crucial to the structure and function of blood vessels. Its deficiency can therefore compromise vascular integrity and encourage the development of atherosclerosis.

Conclusion

These case studies illustrate the importance of metals in maintaining brain and cardiovascular health. Imbalances, whether zinc or copper deficiencies or iron overload, can seriously compromise physiological mechanisms, increasing the risk of stroke. An in-depth understanding of these metal interactions could guide innovative prevention and treatment strategies to reduce the risk of stroke.

Chapter 6: Therapeutic perspectives

6.1. Bioinorganic approaches to stroke treatment

Inorganic and bioinorganic chemistry offer promising prospects for the development of treatments for stroke. Current approaches are based on several key strategies:

- **Thrombolytic therapies**: The use of thrombolytic agents, such as tissue plasminogen activator (tPA), is based on bioinorganic principles. These agents exploit enzymatic mechanisms to dissolve blood clots, thereby restoring circulation. Research continues to explore analogues of tPA based on metal complexes to improve efficacy and reduce side effects (Huang et al., 2021).

- **Metal-based drugs**: Some drugs use metals as cofactors for their activity. For example, copper and zinc complexes are being investigated for their neuroprotective properties. Studies show that these complexes can reduce inflammation and oxidative stress in neuronal cells, which is crucial after a stroke (Kumar et al., 2020).

- **Nanomedicine**: Nanomaterials, often enriched with metals, are being explored to specifically target damaged cells in the brain. These delivery systems can increase the bioavailability of treatments and minimise undesirable systemic effects. For example, gold and iron nanoparticles are being used for targeted drug delivery applications (Gonzalez et al., 2022).

6.2 Innovations in research into metals and medicines

Current research is focusing on integrating metals into therapeutic strategies to treat stroke, with several promising avenues:

- **Metals as therapeutic agents**: Understanding the biological roles of metals in enzyme regulation and electron transport opens the door to the use of these elements in pharmacology. Studies have shown that the addition of iron to certain treatments can improve the efficiency of oxygen transport in affected tissues (Smith et al., 2019).

- **Research into metal complexes**: Transition complexes, such as ruthenium or platinum, are being explored for their anti-tumour potential and their ability to target specific pathways in cells damaged by stroke. Initial clinical trials have shown promising results in reducing brain damage (Chen et al., 2020).

- **Combinatorial approaches**: Combining traditional therapies with bioinorganic approaches is an area of innovation. For example, the use of anticoagulant drugs combined with agents based on metal complexes can improve clinical outcome after stroke (Li et al., 2021).

6.3. Future treatments and targeted therapies

Advances in bioinorganic research point to promising future developments in targeted stroke therapies:

- **Targeted therapies with nanoparticles**: Metal-functionalised nanoparticles offer opportunities for highly targeted treatments, capable of delivering drugs directly to damaged brain tissue, while reducing undesirable effects on healthy cells (Park et al., 2022).

- **Development of biomarkers**: The use of metals as biomarkers in the diagnosis and monitoring of stroke could enable early, personalised intervention. For example, monitoring zinc and copper levels in the blood could help predict the risk of vascular events (Nguyen et al., 2020).

- **Research into metalloenzymes**: Studies into the metalloenzymes involved in critical biological processes, such as fibrinolysis and coagulation, could lead to innovative therapies that specifically target these enzymes to modulate the vascular response after stroke (Thompson et al., 2021).

Conclusion

Summary of key points discussed

This book explored in depth the crucial role of bioinorganics in biological systems, particularly in the context of stroke and fibrinolysis. We have covered the fundamental concepts of bioinorganics, highlighting the importance of metals and metalloenzymes in the regulation of essential physiological processes. The chapters highlighted the complex mechanisms of coagulation and fibrinolysis, illustrating how metals, such as zinc, iron and copper, interact with biomolecules to influence these vital processes.

We also examined thrombolytic agents and the impact of metal complexes on their efficacy, while discussing the clinical implications of metal imbalance on vascular health. Finally, therapeutic perspectives and innovations in metal and drug research offer promising insights into stroke treatment.

The importance of interdisciplinarity between bioinorganics, chemistry and medicine

Interdisciplinarity between bioinorganics, chemistry and medicine is essential to meet the health challenges associated with metals and vascular diseases. By integrating knowledge from different disciplines, researchers can develop innovative approaches to diagnosing and treating stroke, taking advantage of bioinorganic mechanisms. This holistic approach also promotes a better understanding of the complex interactions within biological systems, which is fundamental to the development of more effective and targeted treatments.

Future Research Perspectives

The avenues for future research in bioinorganics are vast and promising, offering significant potential for broadening our understanding of metal interactions in biological systems. Here are some key areas of research, with particular emphasis on innovative techniques and examples of unresolved interactions.

1. Using XAS Spectroscopy

X-ray Absorption Spectroscopy (XAS) is an advanced technique for exploring interactions between metals and biomolecules.

- **Oxidation State Analysis**: XAS can be used to identify the different oxidation states of metals within biological systems, which is essential for understanding their functionality. For example, in metalloenzymes, variations in oxidation state can directly influence their catalytic activity.

- **Coordination and electronic environment**: This technique also provides information on metal coordination structures and electronic environments, which are crucial for understanding enzyme mechanisms. Studies using XAS could, for example, reveal how copper ions interact with enzyme complexes in the catalysis of redox reactions.

2. Studies on Metalloenzymes and their Interactions

Metalloenzymes, which depend on the presence of metals for their activity, are interesting targets for in-depth research.

- **Interactions with various metal ligands**: Exploring the interactions between metalloenzymes and various metal ligands could offer valuable insights. For example, the interaction between metalloenzymes and iron complexes could be poorly understood in certain pathological contexts, such as neurodegenerative diseases.

- **Example of an Unresolved Interaction**: A relevant example is the case of manganese complexes in superoxide dismutase (SOD). Although the importance of manganese in SOD activity is well established, the precise details of the interaction between manganese ions and the active sites of the enzyme, as well as their impact on inhibition by other metals, remain partially resolved. Studies using XAS could clarify how variations in manganese coordination and oxidation state influence enzyme activity, particularly in the context of oxidative stress.

3. Development of Targeted and Personalised Treatments

Research into bioinorganics could contribute to the development of targeted treatments based on an in-depth understanding of metal interactions.

- **Personalised treatments**: By identifying biomarkers based on metal interactions in specific diseases, it would be possible to design treatments that are better adapted to each patient. For example, analysing zinc levels and their interaction with enzymes involved in wound healing could make it possible to personalise therapeutic approaches for patients suffering from chronic wounds.

- **New thrombolytic agents**: The knowledge acquired about metalloenzymes could also lead to the development of new thrombolytic agents that specifically target metal interactions. For example, metal complexes designed to modulate plasmin activity could open up new avenues for treating ischaemic stroke.

Conclusion

By integrating advanced techniques such as XAS spectroscopy and deepening our understanding of metalloenzymes and their interactions, bioinorganics research is well placed to make significant contributions to healthcare. This research could not only enrich our understanding of fundamental biological mechanisms, but also lead to practical advances in the treatment of vascular disease and other medical conditions.

Chapter 7: Role of prostaglandins in stroke

1. Introduction to Prostaglandins

Prostaglandins are bioactive lipids derived from arachidonic acid, an essential fatty acid. They belong to the eicosanoid family and play a fundamental role as mediators in numerous physiological processes. Their chemical structure allows them to interact with specific receptors on cell membranes, influencing a variety of biological functions, including blood pressure regulation, smooth muscle contraction and modulation of the inflammatory response.

In the context of inflammation, prostaglandins exert both pro- and anti-inflammatory effects, depending on their type and environment. For example, prostaglandin E2 (PGE2) is known to promote inflammation and pain by sensitising nerve endings and increasing blood flow to affected areas. On the other hand, other prostaglandins can attenuate the inflammatory response by promoting the resorption of leukocytes and regulating the production of other mediators.

The involvement of prostaglandins in stroke is particularly significant. During an ischaemic stroke, oxygen deprivation leads to the release of arachidonic acid, which stimulates the synthesis of prostaglandins. This response can contribute to cerebral inflammation, aggravating neuronal damage and impairing neuroprotection. Increased levels of PGE2, for example, can induce cerebral oedema, exacerbate damage and compromise functional recovery.

On the other hand, modulation of prostaglandin production pathways could offer therapeutic avenues. Studies suggest that controlling inflammation with anti-inflammatory agents or specific inhibitors of prostaglandin synthesis could reduce the sequelae of stroke, improve recovery and promote neuronal protection. Understanding the role of prostaglandins in the pathophysiology of stroke is therefore crucial to developing innovative and effective treatment strategies.

2. Tranexamic acid

Mechanism of action

Tranexamic acid is a powerful antifibrinolytic that acts primarily by inhibiting fibrinolysis, the process by which blood clots are broken down. Its mechanism of action is based on several key points:

1. **Plasminogen inhibition**: Tranexamic acid binds competitively to the binding sites on plasminogen, preventing its conversion to plasmin. Plasmin is the enzyme responsible for breaking down fibrin, an essential protein in the formation and stabilisation of blood clots.

2. **Clot stabilisation**: By inhibiting the formation of plasmin, tranexamic acid helps to stabilise blood clots that have formed. This is particularly crucial in clinical situations where clot preservation is necessary to prevent life-threatening haemorrhage.

3. **Action on the coagulation cascade**: In addition to its direct effect on plasminogen, tranexamic acid can also influence other aspects of the coagulation cascade, favouring an environment conducive to haemostasis. For example, it can enhance fibrin formation and improve platelet adhesion.

4. **Anti-inflammatory effects**: Some studies suggest that tranexamic acid also has anti-inflammatory properties, which could help reduce local inflammation around brain lesions, thereby minimising neuronal damage.

Clinical use

Tranexamic acid has been the subject of numerous clinical studies, particularly in the context of haemorrhagic stroke. Trials have demonstrated its effectiveness in reducing mortality and bleeding complications. Here are some key points about its use:

- **Clinical trials**: Studies such as the CRASH-2 trial found that early administration of tranexamic acid to trauma patients reduced mortality due to haemorrhage. Although this trial did not focus solely on stroke, it did provide strong evidence for the efficacy of tranexamic acid in acute bleeding situations.

- **Specific indications**: Tranexamic acid is often recommended for patients with haemorrhagic stroke, particularly when the haemorrhage is identified early. Its use can help stabilise the patient's condition and improve clinical outcomes.

- **Safety and tolerability**: Although generally well tolerated, tranexamic acid may present side effects, in particular gastrointestinal disorders and the risk of thrombosis. Its use must therefore be carefully assessed on a case-by-case basis.

3. Aspirin

Mechanism of action

Aspirin, or acetylsalicylic acid, acts primarily as an anti-platelet aggregation agent by inhibiting the enzyme cyclooxygenase (COX), which is essential for the synthesis of thromboxanes. Thromboxane A2 (TXA2), produced by platelets, is a powerful pro-aggregating agent that promotes platelet aggregation and vasoconstriction. By blocking COX-1, aspirin reduces the production of TXA2, leading to a reduction in platelet aggregation and, consequently, an anticoagulant effect. This effect is particularly beneficial in preventing thrombosis, helping to reduce the risk of ischaemic strokes.

Stroke prevention

Current recommendations emphasise the importance of aspirin in both primary and secondary stroke prevention. For primary prevention, aspirin is often recommended for patients with cardiovascular risk factors, such as hypertension or diabetes, to reduce the risk of ischaemic stroke. For secondary prevention, studies show that aspirin significantly reduces the risk of recurrent stroke in patients who have already suffered an ischaemic stroke or transient ischaemic attack (TIA). However, recommendations must be individualised, as the use of aspirin may be associated with bleeding risks, requiring a careful assessment of the benefits and risks.

4 Interactions between these agents

Potential synergy

The combination of tranexamic acid, prostaglandins and aspirin could offer a beneficial synergy in the treatment of haemorrhagic strokes. Tranexamic acid, by stabilising clots, may prevent haemorrhage from worsening, while aspirin, by inhibiting platelet aggregation, could reduce the risk of new clots forming. At the same time, modulation of prostaglandins could help to balance inflammatory responses and protect brain tissue. Together, these treatments could not only reduce mortality but also improve functional recovery by minimising secondary damage caused by inflammation and hypoxia.

Risks and benefits

However, the combined use of these agents requires careful risk-benefit assessment. Comparative studies have shown that although tranexamic acid and aspirin can significantly reduce bleeding and ischaemic events, their combination may increase the risk of bleeding complications. Clinical trials, such as those examined in systematic reviews, have revealed potential side effects, including aspirin-related gastrointestinal disorders and thromboembolic reactions in the event of imbalances in haemostasis management. A personalised approach, taking into account the clinical characteristics of each patient, is essential to optimise results while minimising risks.

5 Conclusion

Summary of key points

Prostaglandins, tranexamic acid and aspirin play complementary roles in the management of cerebral haemorrhage. Prostaglandins modulate the inflammatory response and may influence neuroprotection, while tranexamic acid stabilises clots by inhibiting fibrinolysis, thereby reducing the risk of worsening haemorrhage. Aspirin acts as an anti-platelet agent, limiting the formation of new clots and helping to prevent ischaemic events. Together, these agents can improve clinical outcomes in patients who have suffered a haemorrhagic stroke.

Future prospects

For the future, further research is crucial to better understand the complex interactions between these agents and their mechanisms of action. The exploration of combined treatment protocols could offer more effective strategies for managing haemorrhagic stroke. In addition, the identification of specific biomarkers could help to personalise treatments according to each patient's profile, thereby optimising clinical outcomes. Finally, assessing the long-term effects of these therapies could enrich our understanding of their impact on patients' functional recovery and quality of life.

Conclusion

This book has explored in depth the role of metals and trace elements in vascular health, highlighting their crucial importance in various biological processes. Metals such as copper, zinc and iron, although present in low concentrations, are essential for the proper functioning of biological systems. They are involved in vital mechanisms ranging from the formation of blood vessels to the regulation of inflammation and cellular metabolism.

We have examined how an imbalance in the levels of these metals, whether deficiency or excess, can have serious consequences for health, increasing the risk of cardiovascular diseases such as atherosclerosis and stroke. Case studies have illustrated the harmful effects of insufficient or excessive intake of metals, highlighting the need for a delicate balance to maintain optimal vascular health.

The importance of interactions between metal complexes and enzymes, as well as the mechanisms of action of thrombolytic agents, were discussed, highlighting future prospects for the development of new treatments based on inorganic chemistry. The integration of nanoparticles and coordination complexes into therapeutic formulations represents a promising avenue for improving the efficacy of existing treatments.

Finally, advances in the field of bioinorganics, combined with an interdisciplinary approach, are opening up new possibilities for the research and development of innovative solutions for vascular health. By deepening our understanding of biochemical mechanisms and exploring the complex interactions between metals and biomolecules, we can hope to contribute to significant advances in the prevention and treatment of vascular disease.

In conclusion, careful attention to the balance of trace elements and metal interactions in the human body is essential to promote better health and prevent vascular complications, underlining the importance of ongoing research in this vital area.

Lexicon

1. **Bioinorganics**: Discipline that studies the role of inorganic elements in biological systems.

2. **Cofactor**: Molecule, often a metal, necessary for the enzymatic activity of a protein.

3. **Enzymatic catalysis**: Process by which enzymes increase the speed of biochemical reactions.

4. **Metal complex**: An assembly of a metal and one or more ligand molecules, which can interact with biomolecules.

5. **Free radicals**: Atoms or molecules with an unpaired electron, often reactive and involved in oxidative stress processes.

6. **Coagulation**: Biological process which allows the formation of blood clots to stop bleeding.

7. **Coagulation factors**: Proteins involved in the coagulation process, often designated by Roman numerals (e.g. factor II, factor X).

8. **Thrombin**: Key enzyme which converts fibrinogen into fibrin.

9. **Fibrinogen**: A soluble protein in plasma which is converted into fibrin during coagulation.

10. **Platelet aggregation**: the process by which platelets come together to form a clot.

11. **Fibrinolysis**: Process by which blood clots are broken down.

12. **Plasminogen**: Inactivated protein which, once activated, becomes plasmin.

13. **Plasmin**: Key enzyme which breaks down the fibrin in the blood clot.

14. **D-dimers**: Fibrin degradation products, indicative of fibrinolytic activity.

15. **Tissue plasminogen activator (tPA)**: Enzyme which converts plasminogen into plasmin.

16. **Thrombolytic agents**: Drugs used to dissolve blood clots.

17. **Thrombolysis**: the process of dissolving a thrombus (clot).

18. **Nanoparticles**: Nanometre-sized particles used to target and deliver drugs.

19. **Trace elements**: Chemical elements present in small quantities in the body, but essential to health.

20. **Endothelial dysfunction**: Impaired function of the endothelium, the cell layer lining the blood vessels.

21. **Oxidative stress**: Imbalance between the production of free radicals and the body's ability to detoxify these products.

22. **Atherosclerosis**: A disease characterised by the accumulation of lipids and deposits in the artery walls, increasing the risk of cardiovascular disease.

23. **Haemochromatosis**: A genetic disease in which the body is overloaded with iron, which can cause organ damage.

24. **Thrombolytic therapy**: Treatment aimed at dissolving blood clots in vessels.

25. **Nanomedicine**: Application of nanotechnology in the medical field, particularly for targeted drug delivery.

26. **Neuroprotectant**: Substance which helps to protect neurons against damage or degeneration.

27. **Biomarkers**: Measurable biological indicators that can reflect the state of health or the progression of a disease.

28. **Aspirin**: Non-steroidal anti-inflammatory drug (NSAID) with anti-platelet aggregation properties.

29. **Anti-platelet agent**: Substance which prevents the aggregation of blood platelets, thereby reducing the formation of clots.

30. **Thromboxane A2 (TXA2)** : Lipid mediator produced by platelets, promoting platelet aggregation and vasoconstriction.

31. **Cyclooxygenase (COX)**: Enzyme involved in the conversion of arachidonic acid into prostaglandins and thromboxanes.

32. **Primary prevention**: Strategies designed to prevent the occurrence of a disease in individuals at risk.

33. **Secondary prevention**: Interventions designed to prevent the recurrence of a disease in patients already affected.

34. **Synergy**: Interaction between several agents that produces an effect greater than the sum of the individual effects.

35. **Haemorrhagic stroke**: Type of stroke caused by a rupture of a blood vessel in the brain.

36. **Haemostasis**: Process by which the body stops bleeding.

37. **Haemorrhagic complications**: Undesirable effects associated with increased bleeding.

38. **Thromboembolic**: Relating to the formation of clots that can move through the circulatory system.

References:

1. Bennett, J. M., et al. (2009). "Metal Ions in Biological Systems." Marcel Dekker.

2. Bertini, I., et al. (2001). "Bioinorganic Chemistry: Inorganic Elements in the Chemistry of Life." Wiley.

3. Björk, I., et al. (2008). "Plasminogen Activation and Fibrinolysis: Basic and Clinical Aspects." Thrombosis and Haemostasis, 100(5), 817-826.

4. Baker, S. K., et al. (2012). "Mechanisms of Fibrinolysis." Cardiovascular Drugs and Therapy, 26(6), 551-560.

5. Cohen, M., et al. (2013). "The Influence of Metal Complexes on Plasmin Activity." Biochemical Journal, 453(2), 221-230.

6. Culotta, V. C., & Kolodny, N. H. (2016). "The Role of Copper in the Function of Superoxide Dismutase." Biochemical Society Transactions, 44(6), 1783-1790.

7. Gonzalez, A. et al. (2022). "Nanoparticle-Mediated Drug Delivery for Stroke Therapy." Nanomedicine: Nanotechnology, Biology and Medicine, 39, 102486.

8. Greish, K., et al. (2009). "Urokinase and its Role in Fibrinolysis." Current Pharmaceutical Biotechnology, 10(3), 294-302.

9. Harrison, J. S., & Arooz, R. (2006). "Iron and Its Role in Biological Systems." Journal of Biological Chemistry, 281(33), 23671-23676.

10. Harris, W. S., et al. (2018). "Copper Deficiency and Cardiovascular Risk." American Journal of Clinical Nutrition, 108(5), 1239-1246.

11. Hoffman, M., et al. 2009. "Coagulation: A New Look at an Old Problem." Thrombosis Research, 123(6), 861-865.

12. Hughes, S., et al. (2017). "The Impact of Metal Ions on Blood Coagulation: A Review." Journal of Biological Inorganic Chemistry, 22(5), 831-843.

13. Huang, R., et al. (2021). "Enhancing Thrombolysis with Metal Complexes: A New Approach." Thrombosis Research, 198, 23-32.

14. Hultin, M. B., et al (2004). "Calcium's Role in Fibrinolysis." Blood Coagulation & Fibrinolysis, 15(3), 207-213.

15. Jiang, X., et al. (2011). "Zinc and the Regulation of Plasminogen Activation." Journal of Biological Chemistry, 286(8), 6480-6490.

16. Khan, M. I., et al. (2016). "Metal Complexes as Enhancers of Thrombolytic Activity." Journal of Inorganic Biochemistry, 160, 22-29.

17. Kühn, J. P., et al. (2017). "Copper's Role in Vascular Health." BioMetals, 30(2), 203-213.

18. Kumar, A., et al. (2015). "Mechanisms of Action of Streptokinase." Journal of Thrombosis and Thrombolysis, 40(3), 284-290.

19. Kumar, A., et al. (2020). "Metal Complexes as Neuroprotective Agents: Implications for Stroke Treatment." European Journal of Medicinal Chemistry, 202, 112537.

20. Li, H., et al. (2021). "Combining Anticoagulants and Metal-Based Agents for Stroke Management." Cardiovascular Drugs and Therapy, 35(2), 215-224.

21. Li, X., et al. (2021). "Combination Therapies in Thrombolysis: The Role of Metal Complexes." Therapeutic Advances in Cardiovascular Disease, 15, 175-189.

22. Maret, W. (2009). "Zinc Coordination, Zinc Fingers, and Zinc Enzymes." Biochimica et Biophysica Acta, 1790(3), 630-639.

23. Maret, W. (2018). "Zinc and Cardiovascular Health." Nutrients, 10(6), 797.

24. Mackman, N., et al. (2009). "Role of Tissue Factor in Hemostasis and Thrombosis." Journal of Thrombosis and Haemostasis, 7(1), 1-7.

25. Mann, K. G., et al. (2006). "The Role of Copper in Hemostasis." Blood Reviews, 20(1), 1-8.

26. Miller, J. L., et al. (2015). "Iron and Cardiovascular Health: A Review." Cardiology in Review, 23(1), 29-35.

27. Nielsen, L. B., et al. 2010. "Regulation of Fibrinolysis: PAI-1 and Beyond." Journal of Thrombosis and Haemostasis, 8(6), 1162-1168.

28. Nguyen, T., et al. (2020). "Zinc and Copper Levels as Biomarkers in Stroke Risk Assessment." Stroke, 51(8), 2234-2240.

29. Park, J., et al. (2022). "Targeted Nanoparticle Therapies for Acute Ischemic Stroke." Frontiers in Neuroscience, 16, 835579.

30. Peters, J. R., et al. (2014). "Thrombolytic Therapy: Current Strategies and Future Directions." European Heart Journal, 35(34), 2368-2377.

31. Ravindran, R., et al. (2020). "Zinc Deficiency and Stroke: A Clinical Correlation." Journal of Neurology, 267(1), 115-120.

32. Rosales, C., et al. (2017). "Metals and Platelet Function: Implications for Thrombolysis." Platelets, 28(2), 126-134.

33. Schwartz, B. S., et al. 2006. "Calcium in Coagulation: Physiological and Pathological Perspectives." Cardiovascular Research, 71(4), 671-679.

34. Singh, P., et al. (2018). "Nanoparticles in Medicine: A Review on Thrombolysis." Nanomedicine: Nanotechnology, Biology and Medicine, 14(3), 843-850.

35. Smith, R. J., et al. (2019). "Iron's Role in Improving Oxygen Transport Post-Stroke." Nature Reviews Cardiology, 16(6), 355-366.

36. Thompson, J., et al. (2021). "Metalloenzymes in Vascular Health: Implications for Future Stroke Treatments." Vascular Pharmacology, 135, 106825.

37. Zhang, Y., et al. (2020). "Metal-Organic Frameworks for Thrombolytic Therapy." Chemical Reviews, 120(10), 5091-5130.

38. Zhou, J., et al. (2019). "Metal Interactions and Their Implications for Cardiovascular Disease." Journal of Nutritional Biochemistry, 66, 45-53.

39. Patel, A., et al. (2016). "Aspirin for Primary Prevention of Cardiovascular Events: A Meta-Analysis." Journal of the American College of Cardiology, 67(12), 1411-1420.

40. Mannucci, P. M., et al. (2016). "Combination Therapy with Tranexamic Acid and Aspirin for the Management of Hemorrhagic Stroke: A Review." European Journal of Neurology, 23(7), 1164-1171.

41. Zhang, Y., et al. (2017). "Effect of Tranexamic Acid on Mortality and Transfusion in Trauma: A Meta-Analysis." British Journal of Surgery, 104(3), 201-211.

42. Sacco, R. L., et al. (2008). "Guidelines for the Prevention of Stroke in Patients with Stroke and Transient Ischemic Attack." Stroke, 39(5), 1647-1652.

43. Rothwell, P. M., et al. (2011). "Effect of Daily Aspirin on Risk of First Vascular Events in Patients with High Cardiovascular Risk." Lancet, 377(9761), 81-91.

44. Fitzgerald, G. A. (2004). "Prostaglandins and Thromboxanes: Basic and Clinical Perspectives." The New England Journal of Medicine, 350(19), 1979-1989.

45. Boehme, A. K., et al. (2017). "Stroke and Aging: The Role of Prostaglandins." Current Neurology and Neuroscience Reports, 17(6), 42.

46. Kumar, V. (2014). "Role of Inflammation in Stroke: Insights from Preclinical Studies." Frontiers in Cellular Neuroscience, 8, 79.

47. Wang, Y., et al. (2018). "Inflammation and Repair in Stroke: The Role of Prostagland.

I want morebooks!

Buy your books fast and straightforward online - at one of world's fastest growing online book stores! Environmentally sound due to Print-on-Demand technologies.

Buy your books online at
www.morebooks.shop

Kaufen Sie Ihre Bücher schnell und unkompliziert online – auf einer der am schnellsten wachsenden Buchhandelsplattformen weltweit! Dank Print-On-Demand umwelt- und ressourcenschonend produziert.

Bücher schneller online kaufen
www.morebooks.shop

info@omniscriptum.com
www.omniscriptum.com

Printed by Books on Demand GmbH, Norderstedt / Germany